# THE FUTURE OF TECH LEADERSHIP

Embracing Transhumanism in Business

Golf Ofuka

gcodecloud GmbH

Copyright © 22.12.2024 gcodecloud GmbH

All rights reserved

The characters and events portrayed in this book are Non-fictitious. Any similarity to real persons, living or dead, is coincidental and not intended by the author.

No part of this book may be reproduced, or stored in a retrieval system, or transmitted in any form or by any means, electronic, mechanical, photocopying, recording, or otherwise, without express written permission of the publisher.

ISBN-13: 9798304619219
ISBN-10: 9798304619219

Cover design by: gcodecloud GmbH
Printed in the United States of America

# Dedication

To The Lord and my family—your unwavering love, support, and belief in me have been the foundation of my resilience and success.

To my team at gcodecloud GmbH and Mega Phonebook Nig—this is for your dedication, passion, and commitment to turning vision into reality.

To every tech leader, entrepreneur, and dreamer navigating the challenges of innovation—may this book inspire you to lead with courage, resilience, and purpose.

This is for those who dare to build, to lead, and to transform the future.

# CONTENTS

Title Page
Copyright
Dedication
Introduction
Chapter 1     1
Chapter 2     8
Chapter 3     15
Chapter 4     22
Chapter 5     28
Chapter 6     34
Chapter 7     41
Chapter 8     48
Chapter 9     55
Chapter 10     62
Acknowledgement     69
About The Author     71

# INTRODUCTION

The future of technology is not just about smarter devices, faster processors, or more connected systems—it's about transforming humanity itself. Transhumanism, the integration of technology with human capabilities, represents one of the most profound shifts in human history. As a tech leader or business innovator, this transformation is not something you can afford to watch passively; it's a wave you must learn to ride.

*The Future of Tech Leadership: Embracing Transhumanism in Business* is your guide to navigating this uncharted territory. It explores how transhumanist technologies—such as artificial intelligence, biotechnology, brain-computer interfaces, and enhanced cognition—are reshaping not only industries but the very definition of leadership.

## Why Transhumanism Matters to Leaders

In today's competitive landscape, innovation and adaptability are essential. But as technology becomes increasingly intertwined with human potential, the scope of leadership is expanding. The leaders of tomorrow will not just manage teams or strategies; they will guide organizations through a world where augmented intelligence, enhanced decision-making, and human-machine collaboration are the norm.

Transhumanism offers immense possibilities for businesses:

- Imagine employees with enhanced cognitive abilities, capable of processing vast amounts of data effortlessly.
- Envision decision-making supported by real-time neural-computer interfaces.
- Consider the opportunities in developing products and services that cater to an enhanced human experience.

Yet, with these opportunities come challenges—ethical dilemmas, societal shifts, and the need for a new kind of leadership. Leading in a transhumanist future will require more than technical expertise or business acumen; it will demand a visionary mindset that embraces change while navigating its complexities.

## The Role of Leadership in a Transhumanist World

Leadership has always been about inspiring others, solving problems, and driving progress. In a transhumanist world, these principles remain, but the tools and contexts evolve dramatically. As we step into this future, leaders must:

- Adapt to a workforce augmented by technology.
- Leverage transhumanist advancements to foster innovation and competitive advantage.
- Address the ethical implications of merging humanity with machines.
- Cultivate resilience and adaptability in the face of rapid change.

This book explores these aspects, offering insights, strategies, and case studies to prepare you for what lies ahead.

## A Personal Journey

As someone deeply immersed in technology and business, I've witnessed firsthand how the lines between humanity and technology are blurring. I've seen the incredible possibilities that transhumanist innovations bring—and the leadership challenges they pose. This book is born from those experiences and the belief that the future demands leaders who are both courageous and informed.

This isn't just a book about technology; it's a book about leading in a world where technology redefines what it means to be human. Whether you're a seasoned executive, a start-up founder, or an aspiring leader, this book will help you embrace transhumanism as an opportunity to lead with vision, adaptability, and purpose.

The future is coming fast. Will you be ready to lead in a world where the boundaries between human and machine blur?

Let's step into the future together. Welcome to the age of transhumanist leadership.

# CHAPTER 1

# UNDERSTANDING TRANSHUMANISM

## The Origins of Transhumanism

Transhumanism, as a philosophical movement, traces its origins back to the early 20th century, when thinkers began to explore the implications of human enhancement through technology. Early proponents, such as Julian Huxley, who coined the term "transhumanism" in 1957, envisioned a future where humanity could transcend its biological limitations. Huxley and others contemplated how advancements in science and technology could enable humans to evolve beyond their current physical and cognitive capabilities. This notion laid the groundwork for what would later become a broader movement advocating for the ethical use of technology to improve the human condition.

The technological advancements of the late 20th and early 21st centuries significantly fueled the transhumanist discourse. Innovations in genetics, nanotechnology, artificial intelligence, and biotechnology began to blur the lines between humans and machines, prompting a reevaluation of what it means to be human. As these technologies emerged, transhumanist thinkers, including figures such as Nick Bostrom and Ray Kurzweil, articulated visions of a future where human enhancement could lead to increased intelligence, longevity, and overall quality of life. This convergence of technology and philosophical inquiry attracted attention from various sectors, including business leaders and tech founders, who recognized the potential for transformative change.

The ethical implications of transhumanism have become a focal point of discussion among scholars, ethicists, and business leaders alike. As the possibilities for enhancement grow, questions arise regarding accessibility, equity, and the moral responsibilities of those who develop and implement these technologies. The dialogue surrounding these issues has prompted a deeper examination of how society can ensure that advancements in human enhancement benefit all, rather than exacerbate existing inequalities. For business leaders, navigating these ethical considerations is crucial, as they play a significant role in shaping the future landscape of innovation and its societal impacts.

As transhumanism gained traction, it also intersected with various cultural and social movements, influencing how individuals perceive their relationship with technology. The rise of the digital age has led to an increasing reliance on technology in daily life, further intertwining human experiences with technological advancements. This cultural shift has inspired entrepreneurs and innovators to explore how transhumanist principles can be integrated into business models, creating products and services that augment human capabilities. The fusion of transhumanism with entrepreneurial spirit represents a unique opportunity for leaders to position their organizations at the forefront of a rapidly evolving technological landscape.

In conclusion, the origins of transhumanism are rooted in a rich interplay of philosophical thought, technological advancement, and ethical inquiry. As tech founders and business leaders engage with this movement, they are challenged to consider not only the potential benefits of enhancing human capabilities but also the broader implications of such advancements. The journey forward involves a careful balance of innovation and responsibility, ensuring that the future of leadership aligns with the ideals of transhumanism while fostering an inclusive and equitable society. Embracing this paradigm shift will be essential for leaders who wish to thrive in a world where the boundaries of humanity continue to expand.

## Key Concepts and Principles

Leadership in the context of transhumanism requires an understanding of several key concepts that redefine traditional business practices. One fundamental principle is the integration of technology into human capabilities. This principle emphasizes the enhancement of cognitive and physical abilities through technological means, such as artificial intelligence, biotechnology, and cybernetics. Leaders must recognize how these advancements can be leveraged to not only improve individual performance but also to create a more efficient and innovative organizational culture. The ability to adapt and embrace these technologies will be crucial for businesses aiming to thrive in a rapidly evolving marketplace.

Another vital concept is the importance of ethical considerations in transhumanist leadership. As technology continues to blur the lines between human and machine, leaders must navigate the moral implications of their decisions. This involves fostering a culture of ethical responsibility, ensuring that technological advancements enhance human well-being rather than compromise it. Business leaders should develop frameworks that prioritize transparency, inclusivity, and accountability, guiding their organizations in the responsible use of emerging technologies. This ethical grounding will help build trust among stakeholders and create a sustainable business model that prioritizes long-term success.

Collaboration is also a key principle in the transhumanist landscape. The convergence of diverse technologies and disciplines necessitates a collaborative approach to leadership. Leaders must cultivate environments that encourage interdisciplinary teamwork and innovation, breaking down silos within their organizations. By fostering partnerships with technologists, ethicists, and other experts, business leaders can better navigate the complexities of transhumanism and harness the collective intelligence of their teams. This collaborative spirit not only enhances problem-solving capabilities but also drives the development of innovative solutions that meet the evolving needs of consumers.

In addition to collaboration, adaptability is a crucial concept for leaders in the transhumanist era. The pace of technological change is unprecedented, and businesses must be agile enough to pivot in response to new developments. This requires a mindset that embraces change and encourages experimentation. Leaders should promote a culture of continuous learning, where employees feel empowered to explore new ideas and technologies without fear of failure. By fostering an adaptive organization, leaders can position their companies to not only survive but thrive amidst disruption.

Finally, the principle of holistic development underscores the future of leadership. Transhumanism encourages leaders to view their teams as multifaceted individuals whose growth extends beyond mere productivity metrics. Leaders should invest in the personal and professional development of their employees, recognizing that well-rounded individuals contribute to a more dynamic and innovative workforce. This holistic approach can include mental health support, skill development, and opportunities for creative expression. By prioritizing the overall well-being of their teams, leaders can cultivate a more engaged and motivated workforce, ultimately driving the success of their organizations in the transhumanist future.

## The Intersection of Technology and Humanity

The intersection of technology and humanity represents a pivotal point in the evolution of leadership, particularly within the context of transhumanism in business. As technology advances at an exponential rate, it is reshaping the very fabric of human experience. This transformation offers both opportunities and challenges for leaders who must navigate the complexities of integrating advanced technologies into their organizations while maintaining a focus on human values and ethical considerations. The ability to harmonize technological innovation with human-centric approaches will be crucial for future leaders.

One of the most significant developments at this intersection is the rise of artificial intelligence and machine learning. These technologies have the potential to augment human decision-making processes, enhance productivity, and drive innovation. However, leaders must be aware of the ethical implications associated with AI deployment, including issues of bias, transparency, and accountability. As organizations increasingly rely on algorithms to inform strategic decisions, leaders must ensure that these technologies are used responsibly and that they complement, rather than replace, human judgment.

Moreover, the integration of biotechnology and neurotechnology into the workplace is transforming how leaders approach talent management and employee well-being. Enhancements such as cognitive augmentation, wearable health devices, and genetic modifications can provide significant benefits, including improved mental acuity and physical performance. However, leaders must address the ethical dilemmas surrounding equity and access to these technologies. Ensuring that all employees have the opportunity to benefit from these advancements is essential for fostering an inclusive organizational culture.

The rise of remote work and digital collaboration tools also highlights the intersection of technology and humanity. As businesses adopt more flexible work arrangements, leaders face the challenge of maintaining team cohesion and organizational culture in a virtual environment. Embracing technology to facilitate communication and collaboration is vital, but leaders must also prioritize the human aspects of connection. Building trust and fostering relationships in a digital landscape requires intentionality and empathy, qualities that are increasingly important for effective leadership.

In conclusion, the intersection of technology and humanity presents unique challenges and opportunities for business leaders. By embracing transhumanism principles, leaders can harness the power of technology to enhance human potential while remaining committed to ethical practices and human values. The future of leadership will require a delicate balance between technological advancement and the preservation of what makes us inherently human, ultimately shaping the trajectory of organizations in an increasingly complex world.

# CHAPTER 2

# THE EVOLUTION OF LEADERSHIP

## Historical Perspectives on Leadership

Leadership has evolved significantly over the centuries, shaped by cultural, social, and technological changes. Early forms of leadership were often rooted in authority and domination, with leaders wielding power through force or hereditary privilege. In ancient civilizations, such as those of Mesopotamia and Egypt, leaders were often seen as divine or semi-divine figures, whose authority was justified by their supposed connection to the gods. This perspective emphasized the role of leaders as protectors and providers, establishing a precedent for leadership as a position of responsibility rather than merely a source of power.

As societies progressed into the classical era, particularly in Greece and Rome, leadership began to incorporate more democratic elements. Philosophers like Plato and Aristotle explored the qualities of effective leaders, emphasizing virtue, wisdom, and the importance of serving the public good. This shift laid the groundwork for modern leadership theories, highlighting the necessity of ethical considerations in leadership roles. The emergence of democratic ideals prompted a re-evaluation of leadership, suggesting that effective leaders should not only possess authority but also be accountable to their constituents.

The industrial revolution marked another significant turning point in leadership paradigms. With the rise of organizations and complex hierarchies, leadership became more formalized and structured. Theories such as scientific management introduced the concept of leadership as a function of efficiency and productivity. Leaders were expected to optimize processes and maximize output, often prioritizing organizational goals over individual needs. This mechanistic view of leadership has persisted in many corporate environments, where metrics and performance indicators dominate decision-making.

In the late 20th century, the rise of globalization and technological advancements introduced new challenges and opportunities for leaders. The emergence of the information age shifted the focus from hierarchical control to collaboration and innovation. Leaders began to adopt more participative and transformational styles, emphasizing the importance of inspiring and empowering their teams. This evolution reflects a growing recognition that effective leadership is not solely about directing others but also about fostering an environment where creativity and adaptability can thrive.

Today, as we stand on the brink of a transhumanist future, the understanding of leadership is poised for yet another transformation. The integration of advanced technologies such as artificial intelligence and biotechnology into business practices is reshaping the landscape of leadership. Leaders are now tasked with navigating ethical dilemmas and societal implications brought about by these innovations. The historical perspectives on leadership provide valuable insights, suggesting that as we embrace transhumanism, the qualities of empathy, ethical reasoning, and visionary thinking will be paramount for leaders in the future. This evolution necessitates a redefinition of leadership roles, where the emphasis shifts from traditional authority to a more inclusive and responsible approach that prioritizes the well-being of both individuals and society as a whole.

## Leadership in the Age of Technology

Leadership in the Age of Technology demands an evolution in mindset and approach, particularly as transhumanism reshapes the landscape of business. As technology advances at an unprecedented pace, leaders are required to adapt not only to new tools and platforms but also to the profound implications these innovations have on human capabilities and organizational dynamics. This era calls for a comprehensive understanding of how technology can augment human potential and drive business success, requiring leaders to be both visionaries and practitioners in integrating these advancements into their strategic frameworks.

At the core of leadership in this age is the need for emotional intelligence and empathy, particularly when navigating the complexities of a technologically augmented workforce. Leaders must recognize that the integration of artificial intelligence, biotechnology, and other transhuman innovations can potentially alter employee roles and expectations. Therefore, effective communication and an inclusive culture are paramount. By fostering an environment where team members feel valued and understood, leaders can mitigate the anxiety surrounding job displacement and promote a collaborative atmosphere that embraces change.

Moreover, the ethical considerations of transhumanism present a unique challenge for leaders. As organizations adopt advanced technologies, they must grapple with questions related to privacy, consent, and the moral implications of enhancements. Leaders are tasked with establishing frameworks that prioritize ethical decision-making while driving innovation. This involves not only compliance with regulations but also cultivating a corporate ethos that champions responsible technology use. By leading with integrity, business leaders can build trust with stakeholders and position their organizations as pioneers in ethical transhuman practices.

In addition to ethical considerations, the rapid pace of technological advancement requires leaders to stay informed and adaptable. Continuous learning and development must be embedded within the organizational culture. Leaders should encourage their teams to engage with emerging technologies, fostering a mindset of curiosity and exploration. This approach not only enhances the skill set of employees but also ensures that organizations remain competitive in a constantly evolving marketplace. By prioritizing adaptability, leaders can navigate the disruptions brought about by technology and harness them for strategic advantage.

Ultimately, leadership in the age of technology is about inspiring a shared vision that integrates human potential with technological advancement. Business leaders must articulate a forward-thinking narrative that highlights the benefits of transhumanism, emphasizing how it can lead to enhanced productivity, creativity, and well-being. By championing innovation and fostering a collaborative environment where technology is viewed as an ally rather than a threat, leaders can guide their organizations into a future where human and machine collaboration thrives, driving success in an ever-changing business landscape.

## The Need for a New Leadership Paradigm

The rapid advancement of technology and the rise of transhumanist philosophies necessitate a reevaluation of traditional leadership paradigms. As businesses increasingly integrate cutting-edge technologies such as artificial intelligence, biotechnology, and enhanced human capabilities, leaders must adapt their strategies to navigate this new landscape. The conventional leadership models, often rooted in hierarchical structures and rigid decision-making processes, are becoming less effective in an era where agility, innovation, and collaborative problem-solving are paramount. A new paradigm of leadership is essential to harness the potential of these emerging technologies and to inspire teams in a rapidly changing environment.

In this context, transhumanism presents a unique opportunity for leaders to reimagine their roles. By embracing the principles of transhumanism, which advocate for the enhancement of human intellect and capabilities through technology, leaders can foster a culture of innovation that encourages experimentation and creativity. This approach not only enhances productivity but also motivates employees to push the boundaries of what is possible. Leaders must shift from being mere managers to becoming visionaries who inspire their teams to explore the intersection of humanity and technology, ultimately driving progress and transformation within their organizations.

Moreover, the ethical implications of transhumanism demand that leaders adopt a more holistic perspective. Issues surrounding privacy, equity, and the potential for technological disparity are critical considerations that cannot be overlooked. Leaders need to engage in thoughtful discourse around these topics, establishing ethical frameworks that guide their decisions and influence company culture. By prioritizing these discussions, leaders can build trust with their teams and stakeholders, ensuring that the integration of technology is approached responsibly and inclusively. This reflective leadership style not only strengthens internal cohesion but also enhances the organization's reputation in a society increasingly wary of technological advancements.

The global business landscape is becoming more interconnected, which further emphasizes the need for a new leadership paradigm. As companies expand across borders, leaders must navigate diverse cultural contexts and varying regulatory environments. A transhumanist leadership model encourages a more adaptable and culturally aware approach, enabling leaders to leverage the strengths of diverse teams. By promoting inclusivity and collaboration, leaders can tap into a wealth of perspectives and ideas, fostering creativity and innovation that transcends geographical boundaries. In this way, the new paradigm not only addresses technological advancements but also enriches the human experience within the global marketplace.

Finally, the evolving nature of work itself calls for a

transformation in leadership style. With the rise of remote work, digital collaboration tools, and the gig economy, leaders must cultivate environments where flexibility and autonomy are prioritized. In a transhumanist framework, leaders can leverage technology to enhance communication and collaboration, creating a more dynamic and responsive organizational culture. By embracing this shift, leaders can empower their teams to take ownership of their work, ultimately driving engagement and satisfaction. This new leadership paradigm not only meets the demands of a changing workforce but also positions organizations to thrive in an increasingly complex and interconnected world.

# CHAPTER 3

# THE ROLE OF TECHNOLOGY IN LEADERSHIP

## Emerging Technologies and Their Impact

Emerging technologies are reshaping the landscape of business and leadership, particularly within the transhumanist framework. Innovations such as artificial intelligence, biotechnology, and advanced robotics are not only transforming operational efficiencies but also redefining the very essence of human capability. Business leaders must understand these technologies to leverage their potential effectively, enhancing their organizations' adaptability and resilience in an ever-evolving marketplace. As these technologies become more integrated into daily operations, they present unique opportunities for leaders to foster a culture of innovation and continuous improvement.

Artificial intelligence stands out as a pivotal technology influencing leadership practices. AI can analyze vast amounts of data, providing insights that were previously unattainable. This capability enables leaders to make informed decisions based on predictive analytics rather than intuition alone. Furthermore, AI-driven tools can automate routine tasks, allowing leaders to focus on strategic initiatives that drive growth. For tech founders, harnessing AI not only enhances productivity but also positions their startups at the forefront of technological advancement, appealing to investors and customers alike.

Biotechnology also plays a critical role in the transhumanist vision, particularly in enhancing human performance and health. Advances in genetic engineering, personalized medicine, and wearable health technologies are empowering individuals to optimize their physical and cognitive abilities. Business leaders must recognize the implications of these advancements, as they can influence workforce dynamics and employee well-being. Companies that prioritize health and enhancement technologies can attract top talent, fostering a competitive edge in the marketplace. As organizations embrace these innovations, leaders have the responsibility to create ethical frameworks that govern their implementation.

The rise of advanced robotics offers another layer of transformation in business operations. As robots become more sophisticated, they are capable of performing complex tasks alongside humans, leading to collaborative environments that enhance productivity. This shift necessitates a rethinking of leadership styles, as leaders must learn to manage human-robot teams effectively. Embracing this technology requires a commitment to continuous learning and adaptation, as leaders must equip their workforce with the skills needed to thrive in a hybrid work environment. The ability to navigate these changes will be a hallmark of successful leadership in the future.

Finally, the integration of these emerging technologies necessitates a new approach to ethical leadership. As businesses adopt advanced technologies, leaders must grapple with questions surrounding privacy, security, and the societal implications of transhumanism. A forward-thinking leader will prioritize transparency and ethical considerations in their decision-making processes, ensuring that the adoption of new technologies aligns with the values of their organization and the broader community. By embracing a responsible approach to innovation, business leaders can build trust with stakeholders and position their organizations for sustainable success in the transhumanist era.

## Artificial Intelligence and Decision-Making

Artificial Intelligence (AI) has emerged as a transformative force in decision-making processes across various sectors, fundamentally altering how leaders approach strategy, operations, and innovation. In the realm of transhumanism, where the integration of technology into the human experience is paramount, AI serves as a vital tool that enhances cognitive capabilities and drives informed decision-making. Leaders in the tech and business sectors must recognize the potential of AI to not only streamline processes but also to augment human judgment, leading to more effective and adaptive strategies in an ever-evolving marketplace.

One of the most significant advantages of AI in decision-making is its ability to process vast amounts of data at unprecedented speeds. This capacity allows organizations to analyze trends, forecast outcomes, and identify opportunities that would be impossible for human analysts to discern in a timely manner. By leveraging machine learning algorithms, businesses can uncover insights from big data, enhancing their understanding of market dynamics and consumer behavior. As a result, decision-makers can base their strategies on data-driven insights rather than intuition alone, leading to more reliable outcomes and reduced risks.

Moreover, AI facilitates enhanced collaboration and communication within organizations. By integrating AI systems that can analyze team dynamics and project performance, leaders can make more informed decisions about resource allocation and team structures. These systems can provide recommendations that take into account individual strengths and weaknesses, fostering an environment of continuous improvement. For transhumanist leaders, this means not only optimizing human potential but also creating a more harmonious and efficient workplace where technology complements human efforts.

However, the integration of AI into decision-making processes is not without its challenges. Ethical considerations surrounding data privacy, algorithmic bias, and the potential for over-reliance on technology must be addressed. Business leaders must cultivate an awareness of these issues and establish frameworks that ensure responsible AI use. This includes promoting transparency in AI algorithms and maintaining human oversight to mitigate risks associated with automated decision-making. By doing so, leaders can harness the power of AI while preserving the ethical standards vital to maintaining trust and integrity within their organizations.

As we look to the future, the relationship between artificial intelligence and decision-making will continue to evolve. Leaders who embrace AI as a partner in their decision-making processes will not only enhance their strategic capabilities but also position their organizations at the forefront of innovation. In the context of transhumanism, the synergistic relationship between humans and technology presents opportunities for unprecedented growth and transformation. By adopting a mindset that values collaboration between human intelligence and artificial intelligence, business leaders can navigate the complexities of the modern landscape and drive their organizations toward success in the age of transhumanism.

## Data-Driven Leadership

Data-driven leadership is emerging as a pivotal approach for tech founders and business leaders in the transhumanism era. As organizations increasingly rely on vast amounts of data to guide their strategies, the ability to analyze and interpret this information effectively has become crucial. Data-driven leadership involves making decisions based on empirical evidence rather than intuition or anecdotal experiences. This method not only enhances decision-making accuracy but also aligns organizational goals with measurable outcomes.

In the context of transhumanism, data-driven leadership takes on new dimensions. As technology evolves, leaders must integrate advanced analytics, artificial intelligence, and machine learning into their decision-making processes. These tools allow for the real-time assessment of various metrics, enabling leaders to identify trends, predict future scenarios, and make informed choices that can significantly impact their organizations. The integration of such technologies also fosters a culture of continuous improvement, where data is regularly reviewed to refine operations and enhance overall performance.

Moreover, a data-driven approach enhances transparency and accountability within organizations. When decisions are based on data, it becomes easier to justify actions and communicate strategies to stakeholders. This transparency builds trust among employees, investors, and customers, as they can see the rationale behind key business decisions. In a transhumanist framework, where ethical considerations and societal impacts are paramount, this level of accountability is essential for maintaining credibility and fostering a positive organizational reputation.

Effective data-driven leadership also encourages collaboration across departments. By breaking down silos and promoting a data-sharing culture, organizations can leverage diverse insights and expertise. This collaborative environment leads to more innovative solutions and a stronger alignment of objectives. In a world increasingly influenced by transhumanist ideals, fostering interdisciplinary collaboration is vital, as it brings together various perspectives to tackle complex challenges and drive meaningful change.

Finally, the future of data-driven leadership will require ongoing education and adaptation. As technologies and data sources evolve, leaders must stay informed about emerging trends and tools. Continuous learning will be essential for leaders to harness the full potential of data analytics and maintain a competitive edge. By embracing a mindset of lifelong learning and adaptability, tech founders and business leaders can effectively navigate the complexities of the transhuman era, ensuring their organizations not only survive but thrive in a rapidly changing landscape.

# CHAPTER 4

# TRANSHUMANISM AND THE FUTURE WORKFORCE

## Human Enhancement Technologies

Human enhancement technologies are rapidly evolving, presenting significant implications for leadership and business practices. These technologies encompass a range of innovations designed to improve human capabilities, including cognitive enhancement, physical augmentation, and emotional regulation. As tech founders and business leaders consider the integration of these advancements, it is essential to understand their potential to revolutionize industries and reshape the workforce. The adoption of these technologies not only offers competitive advantages but also poses ethical and operational challenges that leaders must navigate.

Cognitive enhancement technologies, such as brain-computer interfaces (BCIs) and nootropic substances, are at the forefront of human enhancement. BCIs enable direct communication between the brain and external devices, facilitating improved decision-making and increased productivity. Nootropics, or smart drugs, can enhance memory, focus, and learning capabilities. For business leaders, leveraging these cognitive tools can lead to more innovative problem-solving and strategic thinking. However, the potential for inequality and access disparities raises important questions about who benefits from these advancements and how they can be implemented responsibly in the workplace.

Physical augmentation technologies, including exoskeletons and advanced prosthetics, have the potential to transform workforce dynamics, particularly in industries requiring manual labor. These technologies enhance physical capabilities, reduce injury rates, and improve overall worker efficiency. As businesses integrate such enhancements, leaders must consider the implications for employee roles and job descriptions. The shift towards a more augmented workforce may lead to the need for redefined skills and training programs, ensuring that all employees can adapt to and thrive in an enhanced environment.

Emotional regulation technologies, encompassing wearable devices and AI-driven applications, can significantly impact workplace culture and employee well-being. These tools can monitor stress levels, provide real-time feedback, and offer personalized interventions to improve mental health. By fostering a supportive environment that prioritizes emotional intelligence and resilience, leaders can enhance team dynamics and overall productivity. However, the ethical considerations surrounding data privacy and consent in using these technologies must remain a priority for business leaders to maintain trust and integrity within their organizations.

As human enhancement technologies continue to evolve, business leaders must embrace their potential while remaining vigilant about the associated challenges. The successful integration of these technologies requires a strategic approach that addresses ethical concerns, workforce implications, and the need for inclusive access. By fostering a culture of innovation and adaptability, leaders can position their organizations at the forefront of this transformative era, ultimately shaping a future where human and technological capabilities converge for greater success and societal benefit.

## The Changing Nature of Work

The landscape of work is undergoing a profound transformation, driven by advancements in technology, shifting societal norms, and the emergence of transhumanist ideals. As businesses increasingly integrate artificial intelligence, automation, and biotechnology into their operations, the traditional notions of employment and productivity are being redefined. Tech founders and business leaders must adapt to this evolving environment by understanding the implications of these changes on workforce dynamics, job roles, and organizational structures.

One significant aspect of this transformation is the rise of remote and flexible work arrangements. The COVID-19 pandemic accelerated a trend that was already gaining momentum, with many organizations embracing telecommuting as a viable alternative to conventional office settings. This shift not only provides employees with greater autonomy over their work-life balance but also enables companies to tap into a global talent pool. As transhumanist principles advocate for enhancing human capabilities, businesses must leverage technology to create collaborative virtual environments that foster innovation and productivity.

Furthermore, the integration of advanced technologies is reshaping job roles across various industries. Many tasks that were once performed by humans are now being automated, leading to a decline in certain job categories while simultaneously creating new opportunities in fields such as data analysis, AI development, and biotechnology. Business leaders must be proactive in reskilling and upskilling their workforce to prepare for these changes. Emphasizing lifelong learning and adaptability will be crucial for organizations aiming to thrive in this rapidly evolving landscape.

The concept of work itself is also being reexamined. As transhumanism encourages the augmentation of human capabilities, the focus is shifting from traditional job functions to the value of creativity, emotional intelligence, and critical thinking. Future leaders will need to cultivate environments where these skills are nurtured and rewarded, fostering a culture of innovation. This approach not only enhances employee engagement but also drives organizational success in an increasingly competitive marketplace.

Lastly, ethical considerations surrounding the changing nature of work cannot be overlooked. As technology continues to advance, issues related to job displacement, privacy, and the implications of human enhancement must be addressed. Business leaders are tasked with navigating these complexities while ensuring that their organizations uphold ethical standards and promote social responsibility. By embracing transhumanism in a thoughtful manner, companies can lead the charge in shaping a future of work that benefits both individuals and society as a whole.

## Preparing for a Transhuman Workforce

As the concept of transhumanism gains traction within the business landscape, leaders must begin to prepare for a workforce that integrates advanced technologies and biological enhancements. This preparation entails rethinking traditional management practices and embracing a more fluid understanding of human capabilities. Companies will need to assess their current workforce and identify areas where technology can elevate performance. By fostering a culture that encourages innovation and adaptability, organizations can ensure they are ready to integrate transhuman employees effectively.

One of the first steps leaders should take is investing in education and training programs that focus on emerging technologies and their applications in the workplace. As workers become more technologically augmented, understanding how to collaborate with these enhancements will be crucial. Business leaders should encourage continuous learning and provide resources for employees to develop skills that complement their augmented capabilities. This approach not only prepares the workforce for future demands but also promotes a culture of lifelong learning, which is essential in a rapidly evolving landscape.

Additionally, organizations must reassess their recruitment and retention strategies to attract transhuman talent. Traditional hiring practices may not adequately identify candidates who possess both the necessary technical skills and the ability to work alongside enhanced individuals. Leaders should prioritize diversity in hiring, seeking individuals who bring varying perspectives on technology and human enhancement. Creating an inclusive environment that values both biological and augmented capabilities will enhance team dynamics and foster innovation.

Workplace policies will also need to adapt to accommodate the unique needs of a transhuman workforce. This includes establishing guidelines around the ethical use of technology, privacy concerns, and the rights of augmented employees. Business leaders should engage in open dialogues with their teams to address these issues and develop a framework that promotes responsible enhancement. By prioritizing ethical considerations, companies can build trust among employees and foster a collaborative atmosphere that embraces the potential of transhumanism.

Finally, leaders should focus on creating a vision for the future that integrates transhumanism into the core values of their organizations. This vision should articulate the benefits of a transhuman workforce, emphasizing productivity, innovation, and the potential for improved work-life balance. By clearly communicating this vision, leaders can inspire their teams to embrace the changes ahead and work collaboratively toward a future where human capabilities are amplified through technology. The transition to a transhuman workforce represents both a challenge and an opportunity, and with the right preparation, businesses can thrive in this new era.

# CHAPTER 5

# ETHICAL CONSIDERATIONS IN TRANSHUMAN LEADERSHIP

## The Moral Implications of Enhancement

The moral implications of enhancement in the context of transhumanism raise significant questions for tech founders and business leaders. As advancements in biotechnology, artificial intelligence, and human augmentation continue to evolve, leaders must grapple with the ethical dimensions of enhancing human capabilities. The potential for cognitive, physical, and emotional enhancements suggests a future where the line between human and machine becomes increasingly blurred. This reality necessitates a careful examination of what it means to enhance human beings and the societal consequences that may follow.

One of the primary concerns surrounding enhancement is the issue of equity and access. As enhancements become available, there is a risk that only a select few will have access to these technologies, potentially exacerbating existing social inequalities. For business leaders, this raises the moral question of responsibility. How can companies ensure that their innovations do not contribute to a widening gap between the "enhanced" and the "non-enhanced"? Addressing these concerns requires a commitment to inclusivity in the development and distribution of enhancement technologies, fostering a culture that prioritizes social good alongside profit.

Moreover, the concept of identity plays a crucial role in the discourse on enhancement. As individuals enhance their cognitive and physical abilities, questions about the essence of being human arise. Tech founders must consider how these changes impact individual identity and societal norms. The capacity to redefine human capabilities could lead to shifts in personal and collective identities, resulting in potential conflicts regarding what it means to be human. Business leaders are tasked with navigating these complexities, ensuring that enhancements do not undermine essential human qualities such as empathy, compassion, and community.

The ethical implications extend to the workplace as well. With the rise of enhanced employees, traditional metrics of performance and productivity may shift dramatically. Leaders must consider how to create environments that foster collaboration among both enhanced and non-enhanced individuals. Moreover, the ethical treatment of enhanced employees must be prioritized. Questions surrounding consent, autonomy, and the potential for coercion in enhancement decisions must be addressed to cultivate a fair and just workplace.

Ultimately, the moral implications of enhancement will require ongoing dialogue among tech founders, business leaders, ethicists, and society at large. As the capabilities of enhancement technologies expand, a proactive approach to ethics in innovation is essential. By embracing a framework that prioritizes ethical considerations, leaders can contribute to a future where enhancement is leveraged responsibly, ensuring that advancements benefit all of humanity rather than a privileged few. The challenge lies in striking a balance between innovation and ethical responsibility, paving the way for a more equitable and humane future in the realm of transhumanism.

## Balancing Innovation with Responsibility

In the rapidly evolving landscape of technology and business, the drive for innovation often collides with the imperative for ethical responsibility. As tech founders and business leaders, the challenge lies in navigating this delicate balance. While pushing the boundaries of what is possible, leaders must also consider the social, ethical, and environmental implications of their innovations. This dual focus is essential not only for fostering public trust but also for ensuring the long-term sustainability of their ventures in a world increasingly shaped by transhumanist ideals.

Transhumanism advocates for enhancing the human condition through advanced technologies, including artificial intelligence, biotechnology, and robotics. However, the excitement surrounding these advancements can overshadow critical discussions about their potential consequences. Leaders in the tech industry must take a proactive stance in addressing issues such as privacy, data security, and the displacement of jobs. By anticipating the effects of their innovations on society, they can create frameworks that promote responsible usage while still encouraging groundbreaking developments.

A key aspect of balancing innovation with responsibility is fostering an inclusive dialogue among stakeholders. Engaging with diverse perspectives—ranging from ethicists and sociologists to consumers and regulators—can provide invaluable insights into the societal implications of new technologies. This collaborative approach not only enriches the decision-making process but also helps in identifying potential risks and ethical dilemmas early on. By involving various voices in the conversation, tech leaders can better align their innovations with the values and expectations of the communities they serve.

Moreover, business leaders must adopt a long-term vision that prioritizes sustainability over short-term gains. This involves integrating ethical considerations into the core of their business strategies and ensuring that innovation serves humanity rather than endangers it. For instance, developing AI systems that prioritize fairness and inclusivity can mitigate biases that may arise from algorithmic decision-making. By championing responsible innovation, leaders not only enhance their brand reputation but also pave the way for a new business paradigm that values ethical leadership alongside technological advancement.

Ultimately, the future of leadership in a transhumanist context demands a commitment to both innovation and responsibility. As the lines between humans and technology blur, leaders must remain vigilant in their quest to harness the power of technology while safeguarding the human experience. By embracing this dual responsibility, tech founders and business leaders can drive meaningful change, ensuring that their innovations contribute positively to society and elevate the human condition in a sustainable manner.

## Developing an Ethical Framework

Developing an ethical framework in the context of transhumanism is critical for tech founders and business leaders who are navigating the complexities of integrating advanced technologies into society. As innovations like artificial intelligence, biotechnology, and neuroenhancement emerge, the potential benefits come with significant ethical challenges. A well-defined ethical framework not only guides decision-making but also fosters trust among stakeholders, including employees, customers, and the broader community. It is essential for leaders to establish principles that prioritize human dignity, enhance societal well-being, and promote equitable access to technological advancements.

One of the foundational elements of an ethical framework is the principle of transparency. Leaders must commit to openly communicating the intentions and potential impacts of their technologies. This includes providing clear information about how data is collected, used, and protected, as well as the implications of any enhancements or modifications made to human capabilities. By fostering a culture of transparency, businesses can mitigate fears and resistance from the public, thereby creating a more receptive environment for innovation. Moreover, transparency helps to build accountability, ensuring that leaders remain answerable for their actions and decisions.

Another crucial aspect is inclusivity. As transhumanist technologies have the potential to create disparities in access and capability, it is imperative for leaders to prioritize equitable solutions. This means actively engaging diverse voices in the development process and ensuring that the benefits of technology are accessible to all, regardless of socioeconomic status, ethnicity, or geographic location. An inclusive approach not only enhances creativity and innovation by incorporating different perspectives but also fosters a sense of shared ownership in technological advancements, reducing the likelihood of societal backlash against perceived elitism.

Moreover, the ethical framework should emphasize the importance of sustainability. As leaders develop and implement transhumanist technologies, they must consider the long-term impact on both individuals and the environment. This involves assessing the ecological footprint of new technologies and ensuring that their deployment does not compromise the planet's health or future generations' ability to thrive. By prioritizing sustainability, business leaders can position themselves as responsible stewards of both technological progress and environmental preservation, appealing to a growing demographic of conscious consumers and investors.

Finally, a robust ethical framework requires ongoing reflection and adaptation. The rapid pace of technological advancement necessitates that leaders remain vigilant about the evolving ethical implications of their work. This involves engaging with ethicists, stakeholders, and the community to reassess and refine ethical guidelines regularly. Leaders should create forums for dialogue, allowing for the examination of real-world implications and unintended consequences of technologies. By committing to continuous improvement, tech founders and business leaders can ensure that their ethical frameworks remain relevant and effective, ultimately leading to a more responsible and inclusive future in the realm of transhumanism.

# CHAPTER 6

# CASE STUDIES OF TRANSHUMAN LEADERSHIP

## Successful Tech Founders Embracing Transhumanism

In recent years, a growing number of successful tech founders have begun to embrace transhumanism, a movement that advocates for the enhancement of the human condition through advanced technologies. This shift is not just a philosophical stance but a practical approach that many leaders are adopting to redefine the boundaries of human capability and enhance productivity within their organizations. Prominent figures in the tech industry are not only investing in cutting-edge technologies but are also actively exploring how these innovations can lead to significant improvements in human performance, health, and longevity.

One notable example of a tech founder embracing transhumanism is Elon Musk, who has made headlines with his ventures into neural interfaces through his company Neuralink. Musk envisions a future where humans can merge with artificial intelligence, enhancing cognitive abilities and addressing neurological issues. His commitment to integrating technology with human biology exemplifies the potential of transhumanist ideas to transform industries and society as a whole. Such initiatives push the boundaries of what is possible, encouraging other leaders to consider how they can leverage similar advancements for their own organizations.

Another influential figure is Ray Kurzweil, a futurist and co-founder of Singularity University. Kurzweil's predictions about the future of technology and its impact on human evolution have inspired many tech founders to consider how they can contribute to this transformative era. His belief in the power of biotechnology and nanotechnology to enhance human capabilities resonates with leaders who see these innovations as pathways to not only improve employee performance but also to create a healthier workforce. By investing in these technologies, business leaders can foster an environment that prioritizes well-being and productivity.

Moreover, the rise of health and wellness startups, many founded by tech entrepreneurs, illustrates the growing intersection between technology and human enhancement. Companies focusing on personalized medicine, genetic engineering, and mental health solutions are gaining traction as they align with transhumanist principles. These founders understand that enhancing human potential is not just about technological advancement; it also involves nurturing the physical and mental well-being of individuals. This holistic approach can lead to improved organizational culture and employee satisfaction, ultimately driving business success.

As transhumanism continues to gain momentum within the tech community, it presents both opportunities and challenges for business leaders. Embracing this ideology requires a commitment to ethical considerations surrounding human enhancement technologies. Leaders must navigate the complexities of privacy, consent, and potential societal impacts while fostering an innovative culture that prioritizes human advancement. By proactively engaging with these issues, tech founders can position their organizations at the forefront of the transhumanist movement, driving both business growth and societal progress.

## Lessons from Businesses Adopting Advanced Technologies

The adoption of advanced technologies within businesses has provided numerous lessons that can guide leaders in navigating the complexities of a rapidly evolving landscape. One prominent lesson is the importance of agility and adaptability. Companies that have successfully integrated technologies such as artificial intelligence, machine learning, and automation into their operations demonstrate a willingness to pivot and adapt to new tools. This agility not only enhances operational efficiency but also positions organizations to respond quickly to market changes and customer demands. For tech founders and business leaders, embracing a mindset of continuous learning and flexibility is crucial for maintaining a competitive edge in an increasingly transhuman-driven market.

Another significant lesson is the necessity of fostering a culture of innovation. Organizations that prioritize innovation as a core value tend to attract talent that is eager to experiment with advanced technologies. This culture encourages employees to think creatively and take calculated risks, which can lead to groundbreaking solutions and products. Business leaders should focus on creating environments that support collaboration and knowledge-sharing, empowering teams to explore new ideas without the fear of failure. By nurturing a culture that embraces innovation, companies can leverage the full potential of advanced technologies to drive growth and transformation.

The integration of advanced technologies also highlights the importance of ethical considerations in business practices. As companies adopt technologies that enhance human capabilities or alter traditional roles, leaders must be vigilant about the ethical implications of their choices. Issues such as data privacy, algorithmic bias, and the impact of automation on employment require thoughtful deliberation and proactive strategies. Leaders in the transhuman space must prioritize ethical frameworks that guide the implementation of technologies, ensuring that advancements benefit society as a whole and do not exacerbate inequalities. This commitment to ethical leadership will not only build trust with stakeholders but also foster a responsible approach to innovation.

Moreover, successful businesses demonstrate the value of strategic partnerships and collaborations. The complexity of advanced technologies often necessitates expertise beyond an organization's internal capabilities. Forming alliances with tech firms, academic institutions, and research organizations can provide access to cutting-edge resources and knowledge. Such collaborations enable businesses to stay ahead of technological trends and share the risks and rewards associated with innovation. Tech founders and business leaders should actively seek out and cultivate partnerships that align with their strategic goals, thereby enhancing their capacity to leverage advanced technologies effectively.

Finally, the journey of adopting advanced technologies emphasizes the significance of data-driven decision-making. Companies that harness the power of data analytics to inform their strategies are better positioned to understand market dynamics and customer preferences. This insight can lead to more informed decisions regarding product development, marketing strategies, and operational improvements. Leaders must invest in data infrastructure and analytics capabilities, ensuring that their organizations can translate data into actionable insights. By fostering a data-driven culture, businesses can enhance their agility, innovation, and overall effectiveness in a world increasingly shaped by transhumanism.

## Key Takeaways from Industry Leaders

Industry leaders in the realm of transhumanism have shared valuable insights that highlight the necessity for adaptation and foresight in business practices. These leaders emphasize the importance of embracing technological advancements, particularly in the areas of artificial intelligence, biotechnology, and human enhancement. They argue that the future of leadership will require a deep understanding of how these technologies can be integrated into company strategies to foster innovation and competitive advantage. The key takeaway here is that leaders must not only stay informed about technological trends but also actively engage with them to ensure their organizations are positioned for success.

Another significant point raised by industry leaders is the importance of ethical considerations in the implementation of transhumanist technologies. They stress that as businesses adopt advanced technologies, they must also anticipate potential ethical dilemmas and societal impacts. Leaders are encouraged to develop frameworks that prioritize ethical decision-making, ensuring that their innovations do not compromise individual rights or societal values. This proactive approach to ethics can enhance a company's reputation and build trust among stakeholders, which is crucial in an era where consumer awareness is heightened.

Collaboration is also a recurring theme in discussions among industry leaders. They highlight that the complexities of transhumanism require a multi-disciplinary approach that brings together diverse expertise from various fields, including technology, healthcare, and ethics. Establishing partnerships with academic institutions, research organizations, and other businesses can lead to innovative solutions and a more robust understanding of how to leverage transhumanist advancements effectively. Leaders are encouraged to foster a culture of collaboration within their organizations, breaking down silos and promoting cross-functional teamwork to harness the full potential of these technologies.

Furthermore, industry leaders advocate for a forward-thinking mindset that embraces continuous learning and adaptability. The rapid pace of technological change necessitates that leaders remain agile and open to new ideas. They recommend investing in training and development programs that equip employees with the skills needed to navigate and thrive in a transhumanist landscape. By cultivating an environment that values curiosity and adaptability, organizations can better respond to emerging challenges and seize opportunities as they arise.

Finally, leaders are reminded of the importance of vision in navigating the future shaped by transhumanism. They argue that having a clear and compelling vision can guide organizations through uncertainty and inspire teams to work towards common goals. This vision should encompass not only business objectives but also a commitment to enhancing human potential and well-being through technology. By articulating a vision that aligns with the values of transhumanism, leaders can motivate their workforce and create a sense of purpose that resonates with both employees and customers alike.

# CHAPTER 7

# IMPLEMENTING TRANSHUMAN LEADERSHIP STRATEGIES

## Creating a Vision for the Future

Creating a vision for the future is a crucial step for tech founders and business leaders, especially in the context of transhumanism. This philosophical and cultural movement advocates for the enhancement of the human condition through advanced technologies. As leaders in this space, it is essential to envision how these enhancements can reshape industries, transform workplaces, and redefine human potential. A clear vision serves as a guiding light, helping to align efforts and motivate teams toward a common goal, while also anticipating the ethical implications of such advancements.

To craft a compelling vision, leaders must begin by understanding the core principles of transhumanism. This involves recognizing the potential of technologies such as artificial intelligence, genetic engineering, and biotechnology to augment human capabilities. By integrating these concepts into their strategic planning, leaders can identify opportunities for innovation that not only enhance productivity but also improve the overall quality of life. This foresight enables businesses to position themselves as pioneers in a rapidly evolving landscape, fostering a culture of continuous improvement and adaptability.

Engaging stakeholders in the vision creation process is also vital. Tech founders and business leaders should actively involve their teams, customers, and investors to gather diverse perspectives on the implications of transhumanism. This collaborative approach not only enriches the vision but also creates a sense of ownership among stakeholders, leading to stronger commitment and enthusiasm toward achieving the objectives. Regular communication and feedback loops are essential for refining the vision, ensuring it remains relevant and inspiring as the technology and societal context evolve.

Moreover, leaders must consider the ethical and societal implications of their vision. Embracing transhumanism involves navigating complex moral landscapes, such as issues of accessibility, privacy, and the potential for inequality. A responsible vision should address these concerns head-on, outlining how the organization will strive to promote equitable access to enhancements and prioritize the well-being of individuals. By doing so, leaders can build trust with their stakeholders and position their companies as ethical trailblazers in the transhumanist movement.

Finally, a vision for the future should be dynamic and adaptable. The pace of technological advancement means that what seems feasible today may become outdated tomorrow. Business leaders must remain vigilant, continuously reassessing their vision in light of new developments and shifting societal needs. This ongoing evaluation process will not only ensure the relevance of the vision but also empower leaders to seize emerging opportunities that align with their goals. By fostering an environment of innovation and adaptability, organizations can lead the charge in shaping a future where the benefits of transhumanism are realized by all.

## Fostering a Transhuman Culture in Organizations

Fostering a transhuman culture in organizations begins with a clear understanding of the principles underlying transhumanism. This philosophy advocates for the enhancement of human capabilities through technology, which can lead to improved performance, creativity, and overall well-being. For tech founders and business leaders, embracing transhumanism means prioritizing innovation and adaptability within their organizations. It requires leaders to create an environment where technological advancements are integrated seamlessly into the workplace, enabling employees to leverage these tools to enhance their skills and productivity.

A key aspect of fostering a transhuman culture is the commitment to continuous learning and development. Organizations should implement training programs that focus not only on current technologies but also on emerging trends such as artificial intelligence, biotechnology, and neurotechnology. By equipping employees with knowledge and skills that align with transhumanist ideals, companies can cultivate a workforce that is prepared to navigate the complexities of a rapidly evolving technological landscape. This proactive approach to learning fosters a culture of curiosity and exploration, essential for driving innovation.

Collaboration and inclusivity are also vital components of a transhuman culture. Leaders should encourage diverse teams that bring together varied perspectives and expertise, fostering an environment where innovative ideas can flourish. By embracing collaboration across departments and disciplines, organizations can harness the collective intelligence of their workforce. This not only enhances problem-solving capabilities but also promotes a sense of belonging among employees, which is crucial for maintaining high morale and engagement in a transhuman workplace.

Moreover, organizations must prioritize ethical considerations in the integration of technology. As transhumanism raises questions about the implications of enhancing human capabilities, businesses need to engage in open dialogues about the ethical ramifications of their technological choices. Establishing a framework for ethical decision-making can help leaders navigate the challenges associated with adopting transhumanist principles. By fostering a culture of transparency and accountability, organizations can build trust with their employees and stakeholders, ensuring that advancements are pursued responsibly.

Finally, the physical and psychological well-being of employees should be at the forefront of fostering a transhuman culture.

Leaders can implement wellness programs that incorporate technology, such as wearables that monitor health metrics or virtual reality experiences that promote mental well-being. By prioritizing the holistic development of their workforce, organizations not only enhance productivity but also create a culture that values human potential. In doing so, they position themselves at the forefront of the transhumanist movement, ready to lead the way into a future where technology and humanity coexist harmoniously.

## Tools and Techniques for Effective Leadership

Effective leadership in the context of transhumanism requires a unique set of tools and techniques that align with the evolving technological landscape and the human experience. As businesses increasingly integrate advanced technologies, leaders must adapt their strategies to foster innovation while maintaining a human-centric approach. This involves leveraging data analytics, artificial intelligence, and collaborative platforms to enhance decision-making and improve team dynamics. By understanding and utilizing these tools, leaders can navigate the complexities of a rapidly changing environment and inspire their teams to embrace new possibilities.

One of the most important tools for effective leadership in a transhumanist framework is data-driven decision making. Leaders can harness big data analytics to gain insights into market trends, employee performance, and customer preferences. This information allows leaders to make informed decisions that can significantly impact business outcomes. Moreover, employing predictive analytics can help anticipate future challenges and opportunities, enabling leaders to strategize proactively. By embedding data into the leadership process, tech founders and business leaders can create a culture of accountability and transparency, which is essential for fostering trust and collaboration within their organizations.

Another critical technique is the use of artificial intelligence to enhance leadership effectiveness. AI can streamline operational processes, automate routine tasks, and provide advanced analytical capabilities. This allows leaders to focus on strategic initiatives rather than getting bogged down in day-to-day management. Furthermore, AI-driven tools can improve communication and collaboration among team members. For instance, AI-powered project management platforms can facilitate real-time updates and feedback, ensuring that all team members remain aligned with organizational goals. By integrating AI into their leadership practices, business leaders can not only improve efficiency but also cultivate an innovative mindset within their teams.

Embracing a collaborative leadership style is also paramount in the transhumanist business landscape. This approach encourages open communication, inclusivity, and shared decision-making, which are vital for promoting creativity and innovation. Leaders should foster an environment where team members feel empowered to voice their ideas and challenge the status quo. Techniques such as brainstorming sessions, cross-functional teams, and regular feedback loops can facilitate collaboration and enhance team cohesion. By prioritizing collaboration, leaders can tap into the diverse perspectives of their teams, leading to more innovative solutions and a stronger organizational culture.

Lastly, focusing on personal development and emotional intelligence is essential for effective leadership in a transhumanist era. As technology continues to shape the workplace, leaders must remain adaptable and resilient. Investing in leadership training programs that emphasize emotional intelligence can equip leaders with the skills necessary to navigate complex interpersonal dynamics and inspire their teams. Techniques such as mindfulness practices and active listening can enhance leaders' self-awareness and empathy, enabling them to connect more deeply with their team members. By prioritizing personal and emotional growth, leaders can build stronger relationships and create a more engaged and motivated workforce, ultimately driving the success of their organizations in the age of transhumanism.

# CHAPTER 8

# THE FUTURE LANDSCAPE OF BUSINESS LEADERSHIP

## Predictions for Leadership in 2030 and Beyond

The landscape of leadership is set to undergo profound transformations by 2030 and beyond, driven by the convergence of technological advancements and evolving societal expectations. As transhumanism gains traction, leaders will need to adapt their approaches to integrate advanced technologies such as artificial intelligence, biotechnology, and virtual reality. This integration will not only enhance decision-making processes but also foster a more inclusive and dynamic workplace environment. Tech founders and business leaders must prepare for a future where the boundaries between human and machine become increasingly blurred, necessitating a reevaluation of what it means to lead effectively.

One significant prediction is the rise of data-driven leadership, where leaders will rely heavily on real-time analytics and AI insights to inform their strategies. This shift will enable leaders to make more informed decisions, anticipate market trends, and respond swiftly to changes. As organizations harness big data, the ability to interpret and act on this information will become a crucial leadership skill. Leaders who can effectively leverage these tools will gain a competitive edge, allowing them to innovate and adapt in an ever-evolving business landscape.

Furthermore, the concept of emotional intelligence will evolve as technology becomes more integrated into daily operations. Leaders will need to cultivate a unique blend of human empathy and technological literacy. This will include understanding the emotional responses of employees to technological changes, such as automation and remote work. As transhumanist ideals promote the enhancement of human capabilities, leaders must also learn to balance the technological enhancements available to their teams with the intrinsic human values that foster a collaborative culture.

Collaboration will take on new dimensions in the leadership paradigm of 2030 and beyond. The rise of decentralized technologies, such as blockchain, will facilitate more transparent and democratic decision-making processes. Leaders will need to embrace a more participatory leadership style, empowering employees to contribute to strategic initiatives and fostering a sense of ownership. This shift will not only enhance employee engagement but also drive innovation, as diverse perspectives are considered in the decision-making process.

Finally, as ethical considerations surrounding transhumanism gain prominence, leaders will be tasked with navigating complex moral landscapes. The implications of enhancing human capabilities through technology will require leaders to engage in thoughtful discourse about equity, accessibility, and the potential risks of technological dependence. Future leaders will need to champion ethical practices while ensuring that advancements benefit society as a whole. This holistic approach to leadership will be essential in building trust and maintaining a sustainable organizational culture in a rapidly changing world.

## Navigating Challenges and Opportunities

Navigating the complexities of transhumanism in business requires a keen understanding of both challenges and opportunities. As technology advances at an unprecedented pace, leaders in the tech industry must grapple with ethical considerations, societal impacts, and the integration of new capabilities into their organizations. One of the primary challenges is the potential for widening inequality, as access to advanced technologies may be limited to those with resources. Leaders must proactively address these disparities by advocating for inclusive practices and ensuring that their innovations are accessible to a broader audience. This approach not only fosters a sense of social responsibility but also opens up new markets and customer bases.

Another significant challenge lies in the rapid pace of technological change, which can overwhelm organizations unprepared for such shifts. Business leaders must cultivate a culture of adaptability and continuous learning within their teams to stay ahead. This entails investing in employee training and development programs focused on emerging technologies and transhumanist principles. By empowering their workforce with the knowledge and skills needed to navigate these changes, leaders can transform potential obstacles into opportunities for innovation and growth. Embracing a mindset of agility allows organizations to pivot swiftly, responding to market demands and technological advancements effectively.

Moreover, the ethical implications of transhumanism present a complex landscape for leaders to navigate. Issues surrounding privacy, data security, and the moral ramifications of human enhancement technologies require thoughtful consideration and proactive policies. Business leaders must engage in transparent dialogues with stakeholders, including employees, customers, and regulatory bodies, to establish ethical frameworks that guide their practices. By prioritizing ethical decision-making, organizations can build trust and credibility, which are essential for long-term success in an increasingly skeptical society.

Collaboration is another avenue through which business leaders can navigate the challenges posed by transhumanism. Partnering with academic institutions, research organizations, and other businesses can lead to innovative solutions and shared knowledge. These collaborations can foster an environment of experimentation and exploration, allowing leaders to test new ideas and technologies while mitigating risks. By forming strategic alliances, organizations can leverage diverse expertise, enhance their capabilities, and accelerate the development of transhumanist initiatives, thus positioning themselves as pioneers in their respective industries.

Finally, the opportunities presented by transhumanism are vast and varied. From enhancing productivity through advanced cognitive technologies to improving employee well-being via bioengineering, the potential benefits are significant. Leaders who embrace these advancements can create competitive advantages that not only drive profitability but also contribute to societal progress. By focusing on the positive aspects of transhumanism and its capacity to address pressing global challenges, such as healthcare and environmental sustainability, business leaders can inspire their teams and stakeholders alike. This forward-thinking approach not only shapes the future of their organizations but also contributes to a broader vision of a technologically advanced and ethically responsible society.

## The Role of Continuous Learning and Adaptation

In the rapidly evolving landscape of business and technology, continuous learning and adaptation have emerged as critical components for effective leadership. As the transhumanism movement advocates for the enhancement of human capabilities through technology, leaders must embrace a mindset of lifelong learning to remain competitive. This involves not only staying informed about technological advancements but also fostering a culture of curiosity and innovation within their organizations. Leaders who prioritize continuous learning are better equipped to navigate the complexities of a transhuman future, where the integration of AI, biotechnology, and data analytics will redefine operational frameworks.

The pace of technological change necessitates that business leaders develop the skills to anticipate and respond to emerging trends. This requires a proactive approach to learning, where leaders actively seek out new knowledge and insights that can inform their decision-making processes. By engaging with thought leaders, attending industry conferences, and participating in workshops, founders and executives can cultivate a deeper understanding of transhumanist principles and how they can be applied to enhance business strategies. The ability to adapt quickly to new technologies and methodologies will distinguish successful leaders from those who remain rooted in traditional paradigms.

Moreover, fostering a learning-oriented culture within organizations is essential for encouraging innovation and adaptability. Leaders must create an environment where employees feel empowered to pursue professional development and explore new ideas. This can be achieved through initiatives such as mentorship programs, training sessions, and collaborative projects that encourage cross-disciplinary learning. By investing in their workforce's continuous education, leaders not only enhance individual competencies but also drive collective organizational growth. This creates a resilient business model capable of thriving amidst the uncertainties of technological transformation.

As businesses increasingly integrate transhumanist technologies, the importance of ethical considerations in continuous learning becomes paramount. Leaders should prioritize understanding the societal implications of advancements such as artificial intelligence, genetic engineering, and human augmentation. By engaging in discussions around ethics and governance, leaders can guide their organizations in navigating moral dilemmas that may arise from these innovations. This commitment to ethical learning not only builds trust with stakeholders but also positions organizations as responsible players in the transhumanist landscape, enhancing their reputation and long-term sustainability.

In conclusion, continuous learning and adaptation are vital for leaders navigating the transhumanist era. By embracing lifelong learning, fostering a culture of innovation, and prioritizing ethical considerations, business leaders can effectively guide their organizations through the complexities of technological advancements. The future of leadership will demand agility, foresight, and an unwavering commitment to growth and adaptation, ensuring that organizations can harness the full potential of transhumanism while contributing positively to society.

# CHAPTER 9

# BUILDING RESILIENCE IN TRANSHUMAN LEADERSHIP

## Overcoming Resistance to Change

Resistance to change is a common phenomenon in organizations, particularly in the context of technological advancements and transhumanist ideologies. As businesses evolve, leaders often encounter pushback from employees who may feel threatened by new technologies or fear the unknown. This resistance can stem from various factors, including a lack of understanding of the changes being proposed, concerns about job security, and the discomfort that accompanies the shift in established processes. To effectively overcome this resistance, leaders must first recognize and address the underlying fears and misconceptions that contribute to it.

Communication plays a pivotal role in overcoming resistance to change. Leaders must articulate a clear vision of the future and how transhumanist innovations will enhance both individual roles and overall organizational performance. Providing comprehensive information about the benefits of new technologies, such as improved efficiency, enhanced capabilities, and potential for personal growth, can help alleviate fears. Regular updates and open forums for discussion also foster an environment where employees feel heard and valued, encouraging them to embrace the changes rather than resist them.

Training and education are crucial components in facilitating a smooth transition during periods of change. By offering targeted training programs that equip employees with the skills necessary to thrive in a transhumanist landscape, leaders can transform uncertainty into confidence. Workshops, seminars, and hands-on experiences with new technologies can demystify the innovations and empower employees to adapt. Furthermore, continuous learning initiatives can instill a culture of adaptability, where employees view change as an opportunity for development rather than a threat.

In addition to communication and training, involving employees in the change process can significantly reduce resistance. When team members are actively engaged in decision-making and implementation, they are more likely to feel ownership over the changes. Leaders should seek input from various levels within the organization, encouraging collaboration and brainstorming sessions that allow employees to voice their concerns and suggestions. This inclusivity not only builds trust but also cultivates a sense of community, where individuals collectively navigate the challenges associated with change.

Lastly, recognizing and rewarding adaptability can reinforce positive attitudes toward change. Leaders should celebrate milestones achieved during the transition and acknowledge the efforts of teams and individuals who demonstrate resilience and openness to new technologies. By creating a culture that values innovation and adaptability, organizations can foster an environment where resistance diminishes, and enthusiasm for the future of leadership in a transhumanist context thrives. This proactive approach not only prepares the organization for inevitable changes but also positions it as a forward-thinking entity ready to harness the full potential of technological advancements.

## Developing Agility and Flexibility

Developing agility and flexibility is essential for business leaders in the rapidly evolving landscape influenced by transhumanism. As technology continues to advance at an unprecedented pace, organizations must adapt to new tools, methodologies, and societal expectations. Agility allows companies to respond quickly to changes in the market, while flexibility enables leaders to reconfigure resources and strategies as necessary. This dual capability is fundamental for maintaining a competitive edge and fostering innovation in an era characterized by constant transformation.

One of the core components of cultivating agility is adopting a mindset that embraces change rather than resists it. Leaders should encourage their teams to view challenges as opportunities for growth and experimentation. This shift in perspective can be achieved through an organizational culture that prioritizes continuous learning, where employees are empowered to take calculated risks and learn from failures. By fostering an environment where adaptability is celebrated, businesses can create a workforce that is more responsive to technological advancements and shifting consumer demands.

In addition to cultural shifts, implementing agile methodologies can significantly enhance an organization's flexibility. Frameworks such as Scrum or Kanban allow teams to break projects into manageable tasks, facilitating iterative progress and rapid adjustments based on feedback. These methodologies encourage collaboration and transparency, enabling leaders to make informed decisions quickly. Furthermore, the use of technology tools that support these frameworks can streamline workflows and enhance communication, making it easier for teams to pivot in response to changing circumstances.

Another critical aspect of developing agility and flexibility is investing in training and development programs that equip employees with the necessary skills to thrive in a transhumanist business environment. As artificial intelligence and automation continue to reshape job functions, leaders must ensure that their workforce is prepared to leverage these technologies effectively. By providing opportunities for upskilling and reskilling, organizations can create a more adaptable talent pool that can navigate the complexities of new technologies and business models.

Finally, embracing collaboration with external partners and

stakeholders can further enhance an organization's agility. In a world where innovation often arises from cross-disciplinary interactions, forming strategic alliances can provide access to new ideas, technologies, and markets. Collaborating with academic institutions, tech startups, and industry consortia can foster a culture of innovation and agility, enabling leaders to stay ahead of trends and capitalize on emerging opportunities. By recognizing the value of external insights and partnerships, business leaders can cultivate a more flexible organization capable of thriving in the transhuman future.

## Cultivating a Growth Mindset

Cultivating a growth mindset is essential for tech founders and business leaders navigating the complexities of a transhuman future. A growth mindset, as defined by psychologist Carol Dweck, is the belief that abilities and intelligence can be developed through dedication and hard work. In the rapidly evolving landscape of technology and transhumanism, leaders who embrace this mindset are better equipped to adapt to change, foster innovation, and inspire their teams. The ability to view challenges as opportunities for growth rather than insurmountable obstacles is crucial in a world where disruption is the norm.

One of the key components of a growth mindset is embracing failure as a stepping stone to success. In the tech industry, failures are often viewed as setbacks, but those with a growth mindset recognize them as valuable learning experiences. This perspective encourages experimentation and risk-taking, which are vital for innovation. Business leaders should create an organizational culture that celebrates trial and error, allowing team members to explore new ideas without the fear of punitive consequences. By normalizing failure, leaders can foster an environment where creativity thrives, leading to groundbreaking advancements in transhuman technologies.

Furthermore, cultivating a growth mindset requires a commitment to continuous learning. Technology evolves at a breakneck pace, and staying ahead of the curve necessitates a dedication to acquiring new knowledge and skills. Business leaders must model this behavior by investing in their own development and encouraging their teams to pursue ongoing education and professional growth. Providing resources such as training programs, workshops, and access to the latest research can empower employees to expand their capabilities. By prioritizing learning, organizations can remain agile and responsive to the demands of an increasingly sophisticated market.

Collaboration is another essential aspect of nurturing a growth mindset within organizations. In the realm of transhumanism, interdisciplinary collaboration can lead to innovative solutions that transcend traditional boundaries. Leaders should encourage cross-functional teams to work together, leveraging diverse perspectives and expertise. This collaborative spirit not only enhances creativity but also fosters a culture of mutual support, where individuals feel valued and empowered to contribute their ideas. By promoting teamwork, leaders can build a strong foundation for innovation that aligns with the principles of transhumanism.

Lastly, self-reflection plays a critical role in cultivating a growth mindset. Leaders should regularly assess their own beliefs, behaviors, and attitudes towards challenges and setbacks. This introspection can reveal areas for improvement and highlight the importance of resilience. By sharing their experiences with their teams, leaders can model vulnerability and authenticity, demonstrating that growth is a continuous journey. Encouraging team members to engage in self-reflection can further strengthen this mindset, leading to a more cohesive and innovative workforce. In the transhuman era, where adaptability and forward-thinking are paramount, cultivating a growth mindset will be instrumental in achieving long-term success.

# CHAPTER 10

# CONCLUSION EMBRACING THE FUTURE OF LEADERSHIP

## The Journey Ahead

The journey ahead for leaders in the realm of transhumanism presents a unique intersection of technology, ethics, and human potential. As advancements in artificial intelligence, biotechnology, and cognitive enhancement continue to reshape the business landscape, leaders must navigate the complexities of integrating these innovations into their organizational strategies. Understanding the implications of transhumanism is not merely about adopting new technologies; it involves a fundamental rethinking of what it means to lead in a world where human capabilities are augmented. Leaders must be prepared to address both the opportunities and challenges that come with these transformative changes.

A pivotal aspect of this journey is the need for a robust ethical framework. As tech founders and business leaders, embracing transhumanism requires an understanding of the moral considerations associated with enhancing human abilities. Questions regarding equity, access, and the potential for creating societal divides must be at the forefront of decision-making processes. Developing policies that prioritize inclusivity and fairness will not only foster a positive organizational culture but also position businesses as responsible pioneers in the transhumanist movement. Leaders must engage in ongoing dialogue with stakeholders to ensure that ethical considerations are woven into the fabric of innovation.

Moreover, the journey ahead will demand a shift in leadership styles. Traditional models may no longer suffice in a landscape characterized by rapid change and uncertainty. Leaders must cultivate adaptive, collaborative environments that encourage experimentation and creativity. By fostering a culture of continuous learning, organizations can empower their teams to embrace change and drive innovation. This includes investing in training programs that enhance both technical skills and emotional intelligence, ensuring that leaders and their teams are equipped to thrive in an increasingly complex world.

As businesses integrate transhumanist principles, the role of technology in augmenting leadership capabilities cannot be overlooked. Tools such as data analytics, AI-driven insights, and enhanced communication platforms will become indispensable in decision-making processes. Leaders must harness these technologies to improve operational efficiency, strengthen customer relationships, and drive strategic growth. Embracing a tech-savvy approach will enable leaders to leverage the full potential of their organizations, creating a competitive advantage in a rapidly evolving market.

Finally, the journey ahead is not solely about technology; it is also about vision. Leaders must articulate a clear and compelling narrative that aligns their organizational goals with the broader aspirations of transhumanism. This vision should inspire and motivate teams while also resonating with customers and stakeholders. As the lines between humans and machines blur, businesses that successfully integrate transhumanist ideals will not only lead in profit but also in purpose. By championing a future that enhances human capabilities, leaders can create a lasting impact that transcends traditional business metrics, paving the way for a more equitable and innovative society.

## Inspiring Future Leaders

Inspiring future leaders in the context of transhumanism involves cultivating a mindset that embraces innovation, ethical considerations, and the continuous evolution of human capabilities. As technology advances at an unprecedented pace, the leaders of tomorrow must not only be adept at navigating this landscape but also be visionary in their approach. They will need to leverage emerging technologies such as artificial intelligence, biotechnology, and cognitive enhancement to enhance organizational efficiency and drive societal change. This requires a commitment to lifelong learning and adaptability, ensuring leaders can pivot in response to new challenges and opportunities.

Central to inspiring future leaders is the emphasis on ethical leadership. As advancements in technology blur the lines between human and machine, leaders must grapple with complex ethical dilemmas. They should be equipped to address questions around privacy, consent, and the implications of cognitive augmentation on workforce dynamics. By fostering a culture of ethical decision-making, leaders can set a precedent for integrity in their organizations. This inspires future leaders to prioritize ethical considerations when developing and implementing transhuman initiatives, ensuring that progress does not come at the expense of fundamental human values.

Moreover, collaboration will be a cornerstone of effective leadership in a transhumanist future. Future leaders must inspire diverse teams, encouraging interdisciplinary collaboration that brings together experts from various fields. This collaborative spirit not only enhances problem-solving capabilities but also fosters a culture of inclusivity and innovation. By creating environments where diverse perspectives are valued, leaders can inspire future generations to think outside the box and approach challenges with a multifaceted lens. This approach will be essential in harnessing the full potential of transhuman technologies, which often require input from both technical and social science domains.

Inspiring future leaders also involves a commitment to sustainability and social responsibility. As technology continues to evolve, the impact of these advancements on society must be carefully considered. Leaders in the transhumanist space should advocate for practices that not only drive business success but also contribute positively to the environment and society at large. By instilling a sense of purpose and responsibility, they can inspire the next generation to view leadership as a platform for creating positive change, rather than merely pursuing profit. This perspective will help shape a future where technological advancements are aligned with the greater good.

Finally, mentorship plays a crucial role in developing future leaders within the transhumanist framework. Established leaders should actively engage in mentoring relationships, sharing their experiences, insights, and lessons learned in navigating the complexities of technology and leadership. This not only helps build a strong foundation of knowledge but also fosters a sense of community among aspiring leaders. By providing guidance and support, mentors can empower future leaders to take risks and innovate, ultimately cultivating a new generation of thinkers who are equipped to lead in an increasingly transhuman world.

## Call to Action for Business Leaders

The landscape of business is rapidly evolving, and as leaders, it is imperative to recognize the potential of transhumanism in driving innovation and growth. Embracing the principles of transhumanism requires a proactive approach from business leaders who must not only adapt to technological changes but also anticipate how these advancements can enhance human capabilities. The integration of technology and human intelligence will redefine competitive advantages, enabling organizations to thrive in an increasingly complex environment.

As technology continues to advance at an unprecedented rate, the role of business leaders is shifting. It is essential to cultivate a mindset that embraces change and innovation. Leaders must invest in understanding emerging technologies such as artificial intelligence, biotechnology, and cybernetics. By fostering a culture of curiosity and experimentation within their organizations, leaders can inspire their teams to explore the possibilities that transhumanism offers, leading to breakthroughs that can propel their companies forward.

Furthermore, collaboration will be a key driver of success in this new era. Business leaders should forge partnerships with tech innovators, researchers, and academic institutions to harness collective knowledge and expertise. These collaborations can lead to the development of cutting-edge solutions that address complex challenges while enhancing human potential. By engaging with a diverse array of stakeholders, leaders can position their organizations as pioneers in the integration of transhumanist concepts into business practices.

Ethical considerations must also be at the forefront of this transformation. As leaders adopt transhumanist principles, they must navigate the ethical implications of enhancing human capabilities through technology. It is crucial to establish frameworks that prioritize the well-being of individuals and society as a whole. By fostering transparency and accountability in their decision-making processes, leaders can build trust among employees, customers, and the broader community, ensuring that the benefits of technological advancements are shared equitably.

In conclusion, the call to action for business leaders is clear: embrace the opportunities presented by transhumanism and lead with vision and responsibility. By investing in technology, fostering collaboration, and prioritizing ethical considerations, leaders can not only transform their organizations but also contribute to a future where human potential is maximized through the thoughtful integration of technology. The time for action is now, and those who lead the way will shape the future of business in profound and lasting ways.

# ACKNOWLEDGEMENT

Writing *The Future of Tech Leadership: Embracing Transhumanism in Business* has been an incredible journey, and it would not have been possible without the support, inspiration, and guidance of many remarkable individuals.

First and foremost, I extend my deepest gratitude to my family, whose unwavering encouragement has been a constant source of strength. Your belief in my vision and your patience through countless hours of research and writing have made this work possible.

To my team at gcodecloud GmbH and Mega Phonebook Nig, thank you for being a daily reminder of the power of collaboration, innovation, and resilience. Your passion for embracing the future of technology fuels my commitment to exploring and sharing ideas that shape our industry.

To the pioneers, thinkers, and innovators in the fields of transhumanism and emerging technologies, I am indebted to your groundbreaking work and visionary insights. Your contributions have laid the foundation for the ideas explored in this book and continue to inspire leaders around the world.

I am also immensely grateful to my mentors and advisors who have guided me throughout my career. Your wisdom and perspective have been instrumental in shaping my understanding of leadership and the role technology plays in driving change.

To the readers and leaders who are open to exploring the possibilities of a transhumanist future, this book is dedicated to you. Your curiosity, courage, and commitment to innovation are what drive progress and transformation.

Finally, to everyone who has been part of my journey—whether through thought-provoking conversations, challenging debates, or unwavering support—thank you. This book is a reflection of the collective effort, inspiration, and shared vision of so many.

As we step into the future, let us lead with purpose, embrace the possibilities of human-technology integration, and create a world that balances innovation with humanity.

— Golf Ofuka

# ABOUT THE AUTHOR

## Golf Ofuka

Golf Ofuka is the dynamic CEO and Founder of gcodecloud GmbH and Mega Phonebook Nig, recognized for his innovative approach to technology and business. His career and entrepreneurial journey highlight his expertise in the software and IT industry, alongside a robust knowledge of business strategy, development, and market positioning. Based in Berlin, Germany, Golf leverages his strong technical foundation and a sharp business acumen to spearhead ventures that bridge technological innovation with market needs.

Background and Leadership
Golf's career path reflects a blend of technology mastery and entrepreneurial insight, making him a valuable leader in today's fast-paced digital landscape. As the CEO and driving force behind gcodecloud GmbH, he leads a team that is transforming software solutions for modern enterprises. His leadership is marked by a deep commitment to fostering innovation within his organization and a clear vision for scaling products and services that meet the evolving demands of global markets.

Similarly, through his leadership at Mega Phonebook Nig, Golf demonstrates his commitment to impactful solutions by connecting businesses and individuals across various communication and data management platforms. His approach emphasizes usability and accessibility, making it easier for clients to integrate these solutions seamlessly into their operations.

www.ingramcontent.com/pod-product-compliance
Lightning Source LLC
Chambersburg PA
CBHW070355230526
45471CB00006B/2579